More Without Words

Mathematical Puzzles to Confound and Delight

James Tanton

TarquinGroup
www.tarquingroup.com

By the same author from Tarquin

Without Words
36 visual puzzles to test and delight you!
ISBN 978 1 907550 23 2
Ebook ISBN 978 1 858118 11 6

Publisher's Note:
We encourage you, before you start, to read the Introduction – these puzzles will often test you and different types of puzzle will test you in different ways. Some you may find easier than others – but if you get really stuck there are some hints and tips on the last three pages.

To check answers and to find out more, go to www.tarquingroup.com and search for Without Words - the link to get the answers and further reading is on the relevant book's page.

© James Tanton 2015
ISBN 978 1 907550 24 9
Ebook ISBN 978 1 858118 12 3
Printed in the UK

tarquin publications
Suite 74, 17 Holywell Hill
St Albans, AL1 1DT, UK
info@tarquingroup.com
www.tarquingroup.com

A FEW BRIEF INFORMAL WORDS (!!)

Ready for another round of wordless puzzles? Here they are!

Following WITHOUT WORDS comes MORE WITHOUT WORDS, a collection of 35 additional immediately accessible but deeply mathematical puzzles designed to confound and delight. As before, not a single word is written.

These puzzles are universal: they transcend the barriers of language and culture, literally, and are thereby accessible to all people on this globe. Moreover, the puzzles themselves speak the universal truth of mathematics.

As before, take your time to mull on each puzzle, be it over hours, days, or weeks. Some of these puzzles are hard and won't be solved in one sitting. Set the stage for epiphanies and flashes of insight to come of their own accord by being patient and gentle with your thinking. Give permission and space for your subconscious to mull. There is never a need to rush in doing true mathematics!

Any relevant insight garnered from a particular puzzle should be deemed a success, even if it does not solve the challenge completely. (For those looking for guidance in case frustration becomes too much to bear, a tips section appears at the end of the book.)

True mathematics "doing" is thrilling, frustrating, compelling, and humbling. It's an intellectual roller coaster. Enjoy the ride!

JST

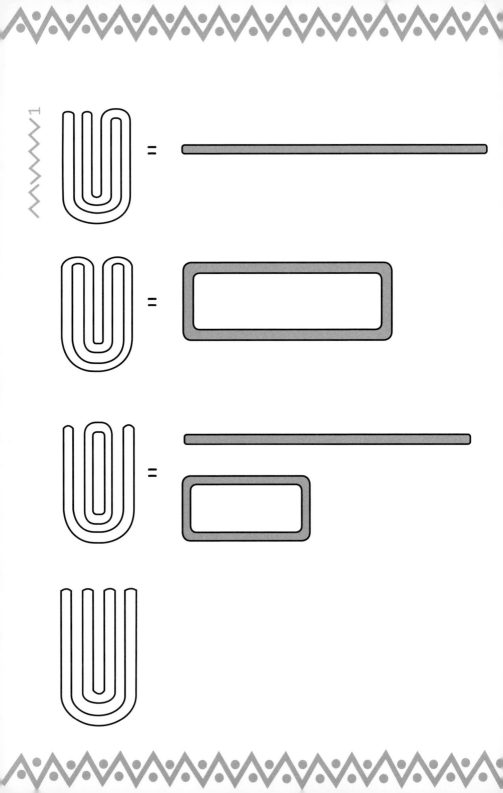

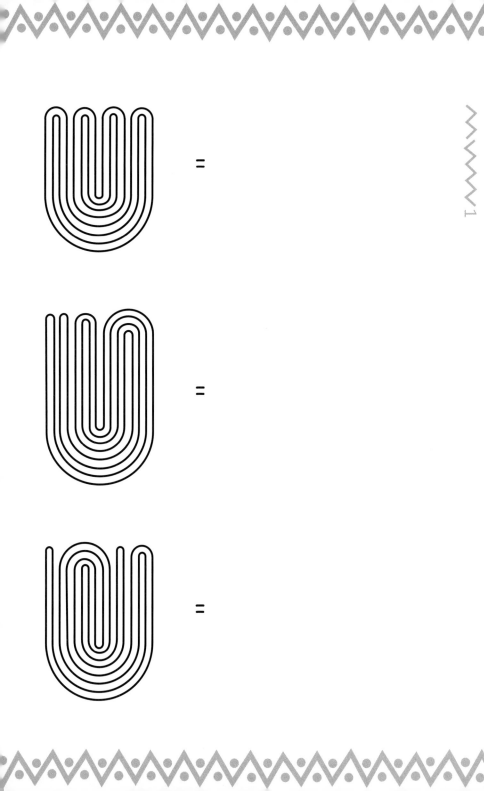

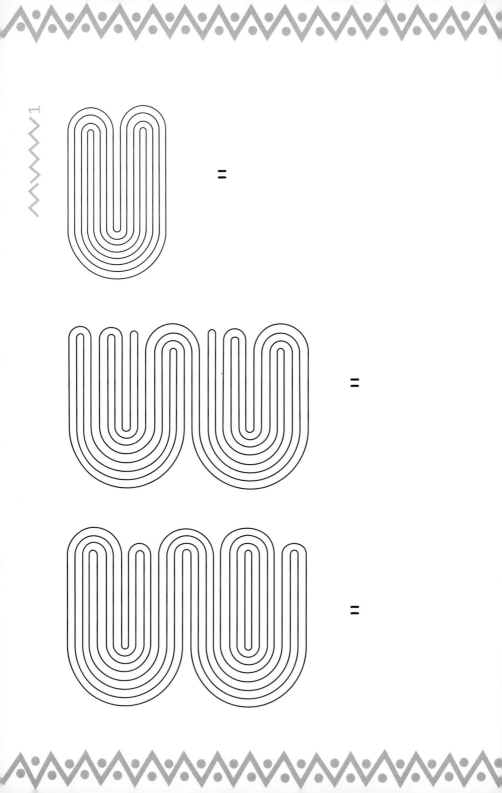

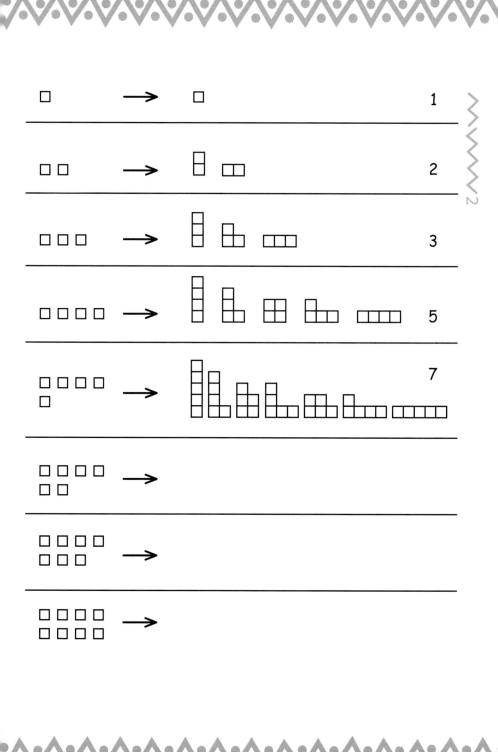

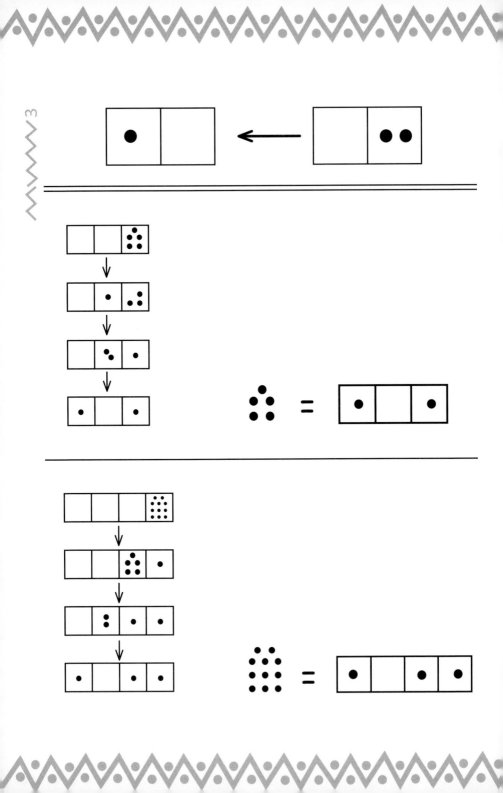

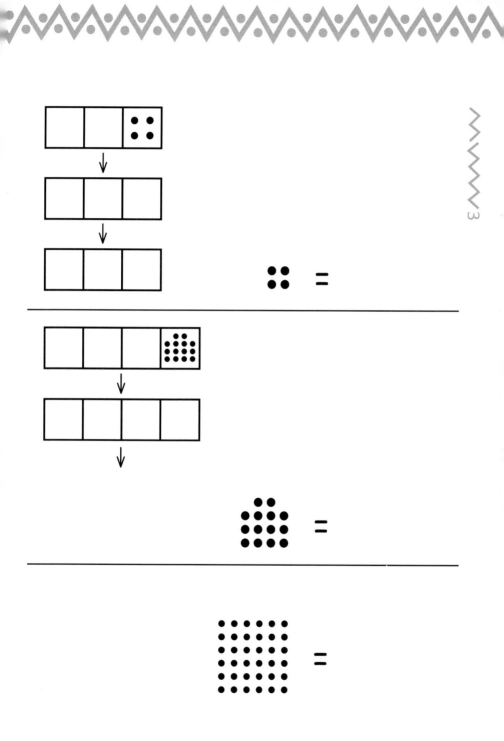

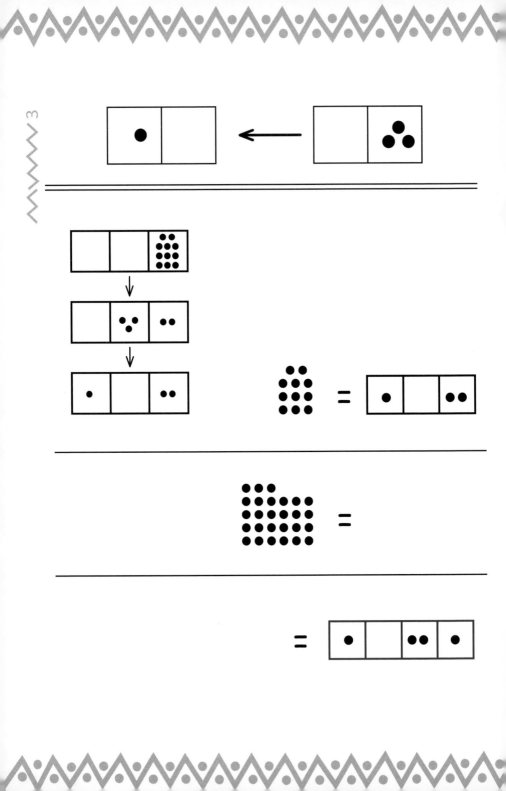

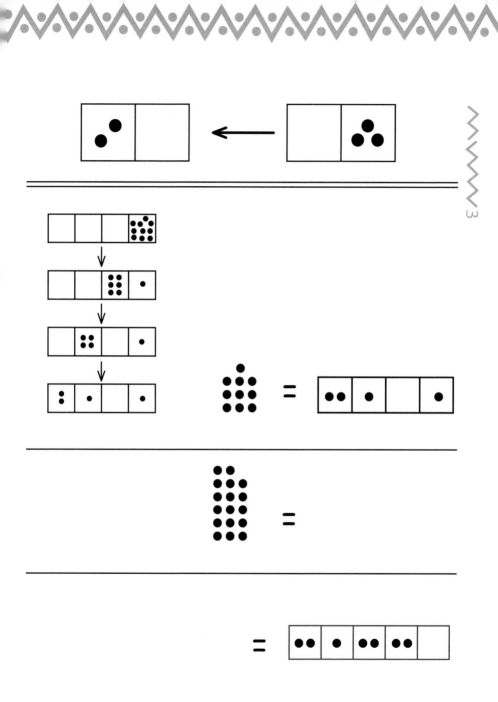

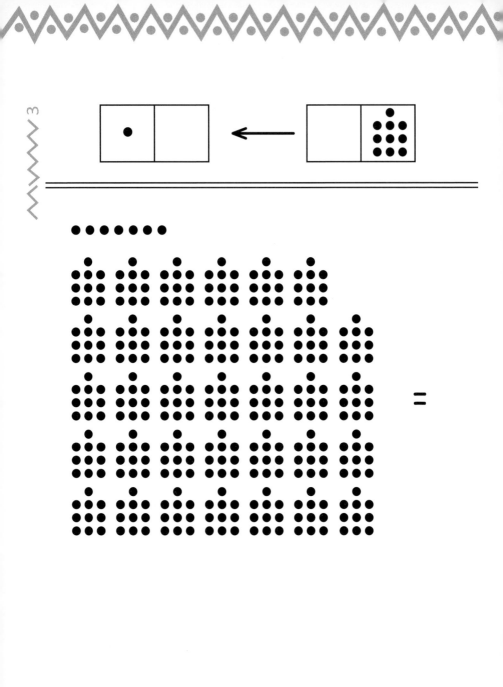

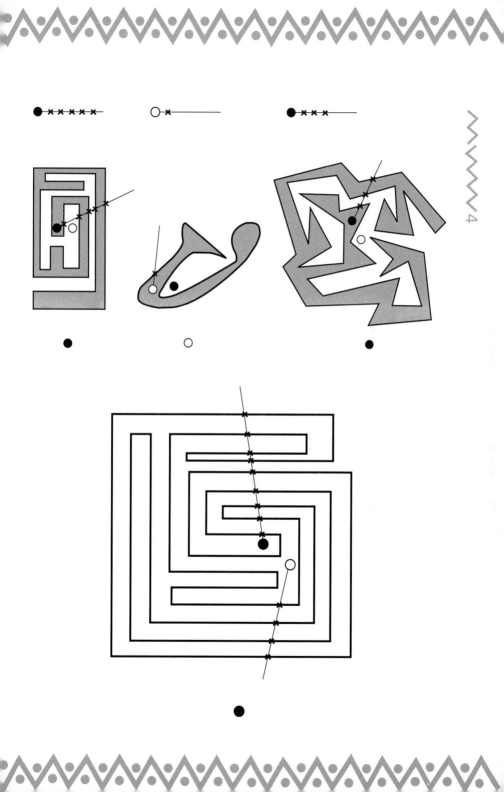

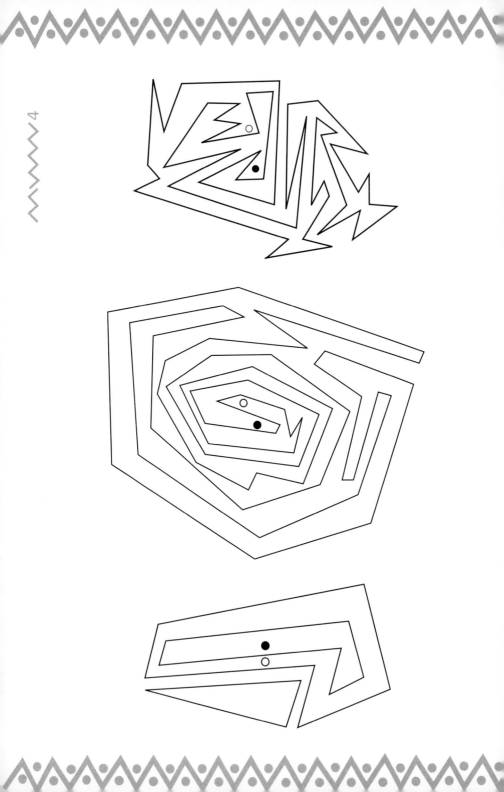

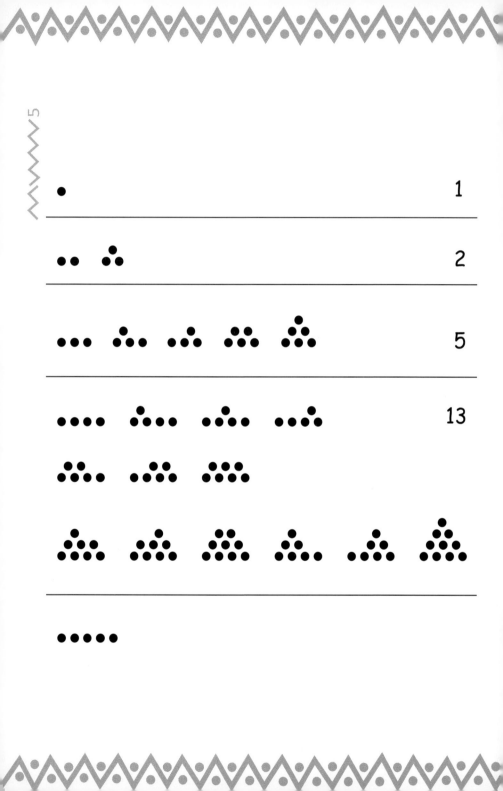

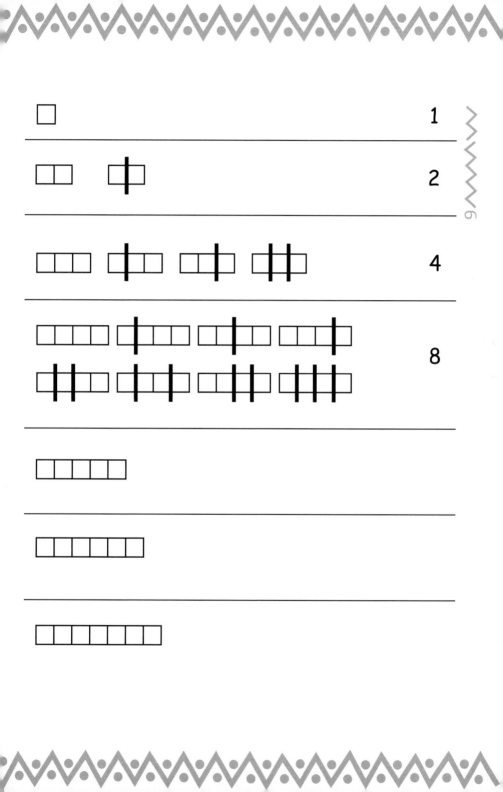

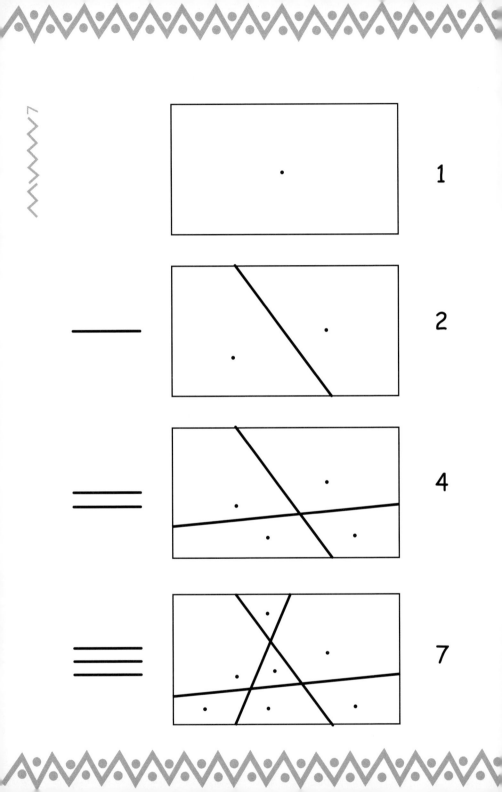

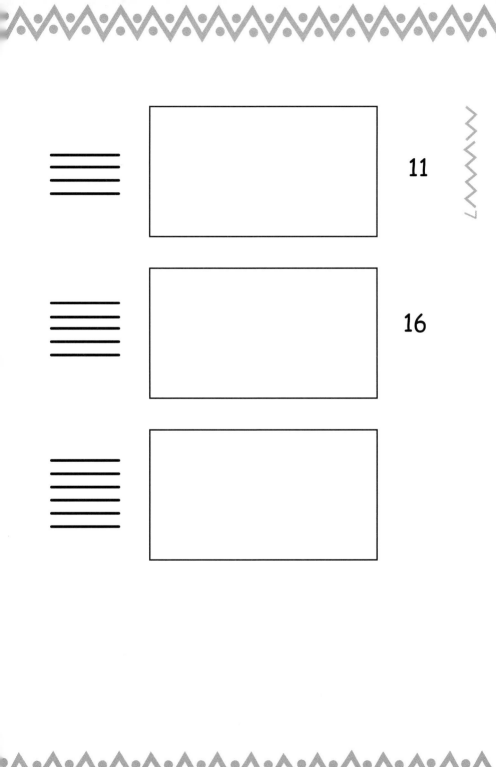

11

16

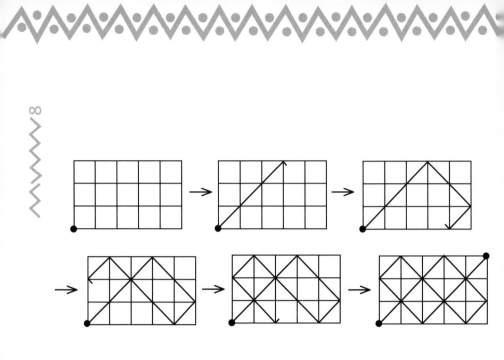

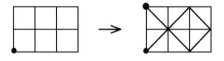

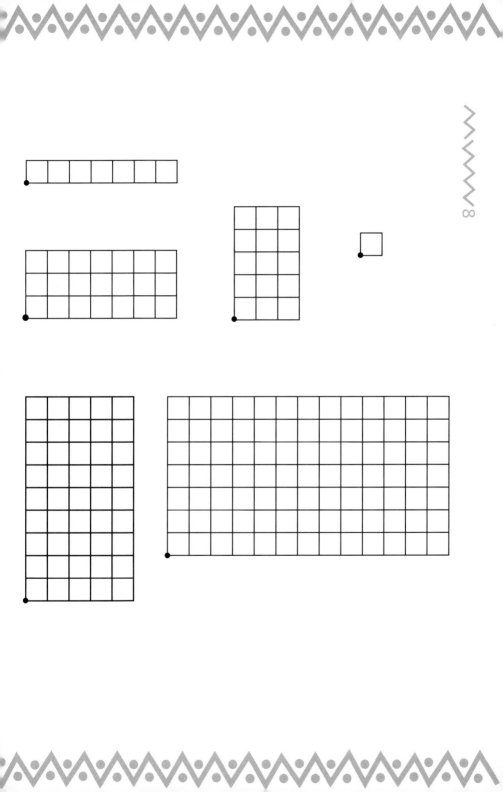

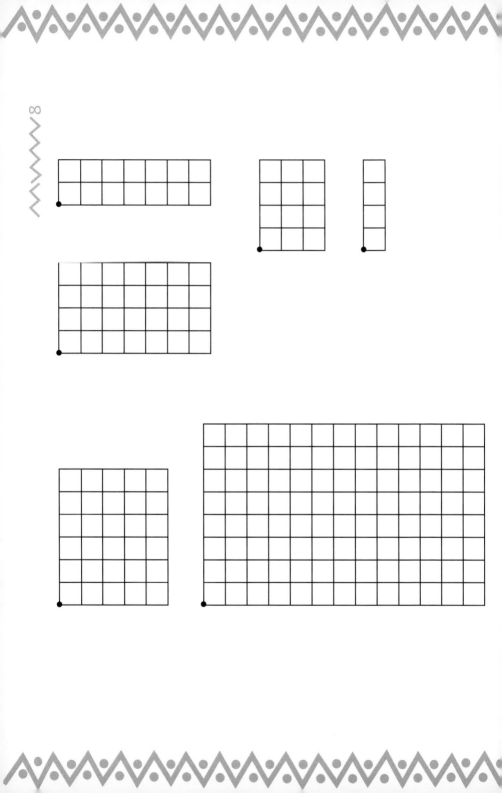

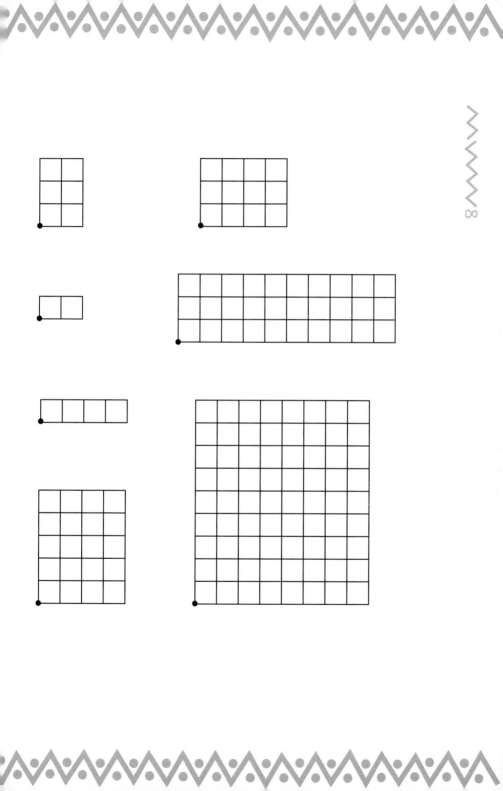

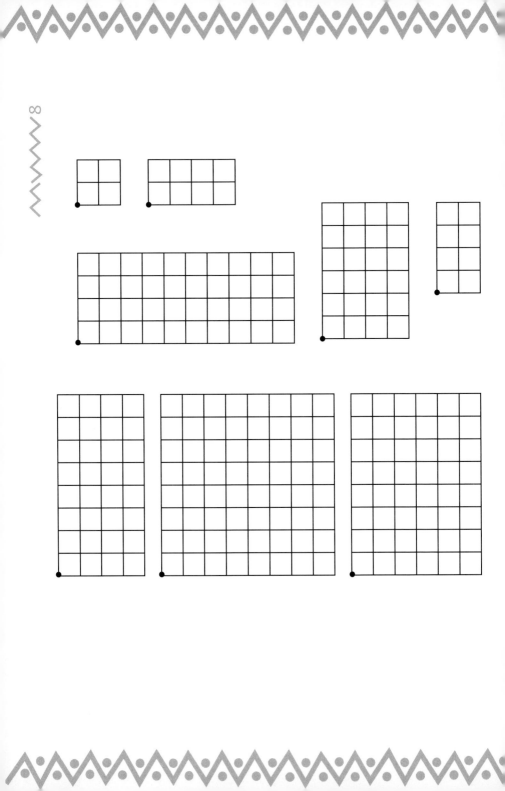

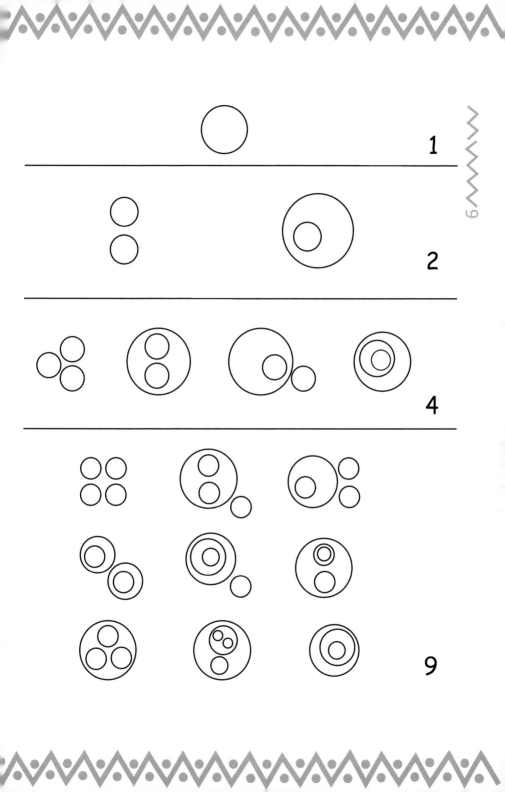

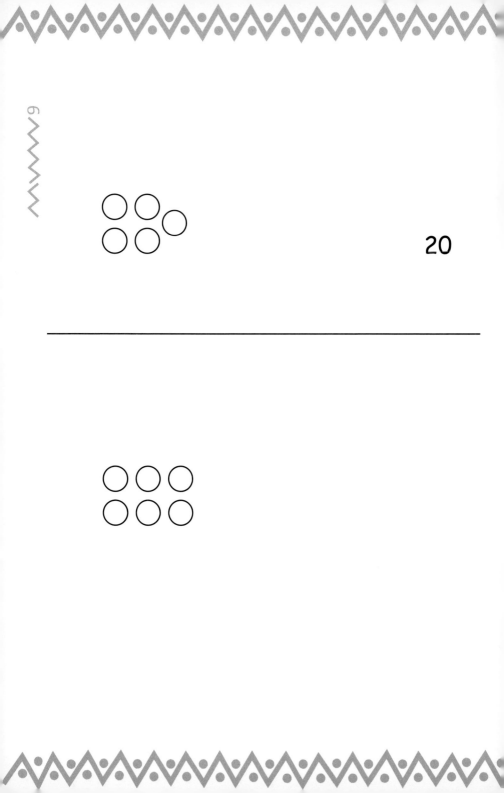

20

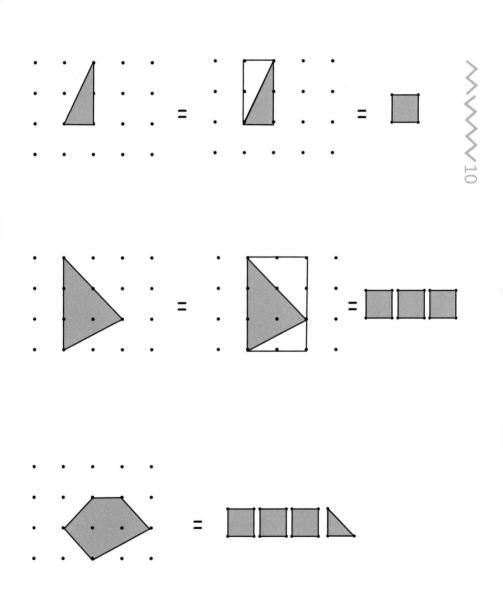

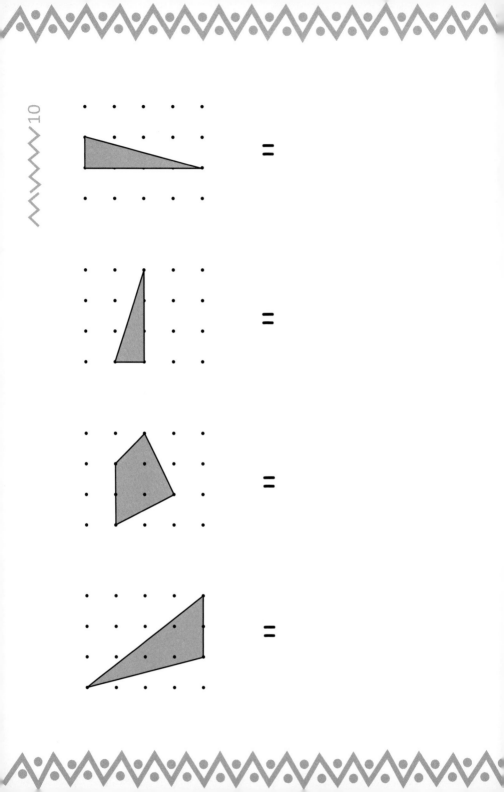

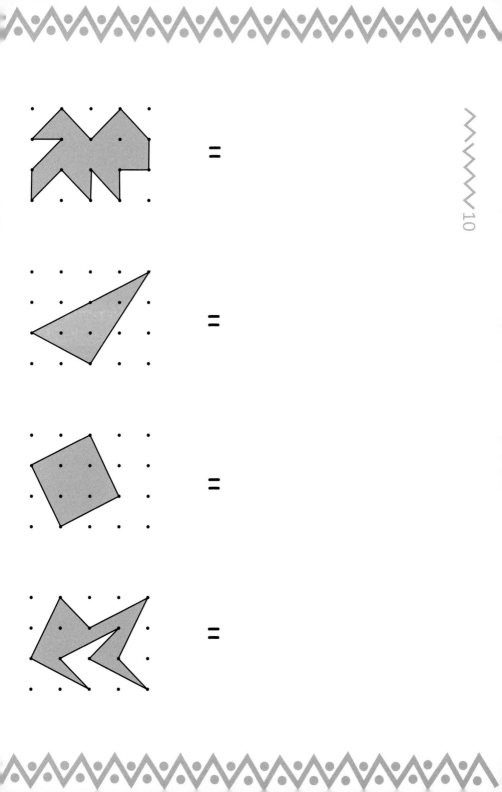

=

=

=

=

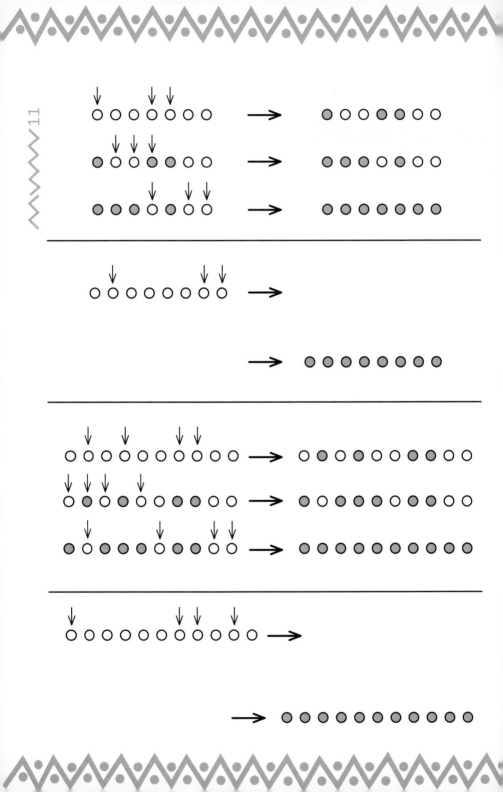

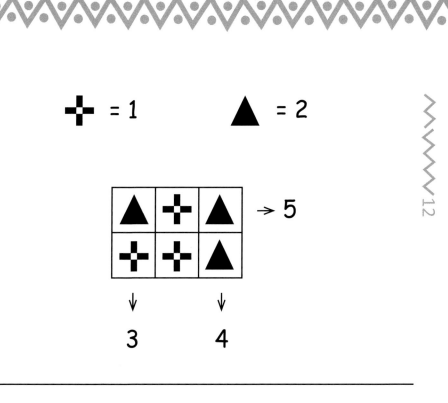

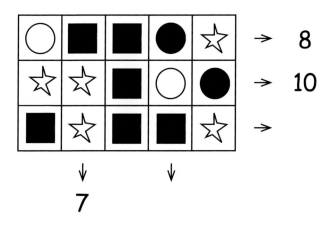

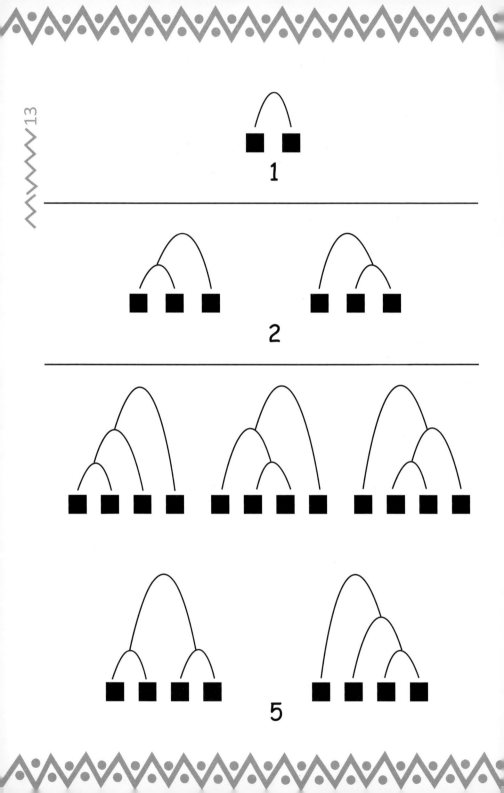

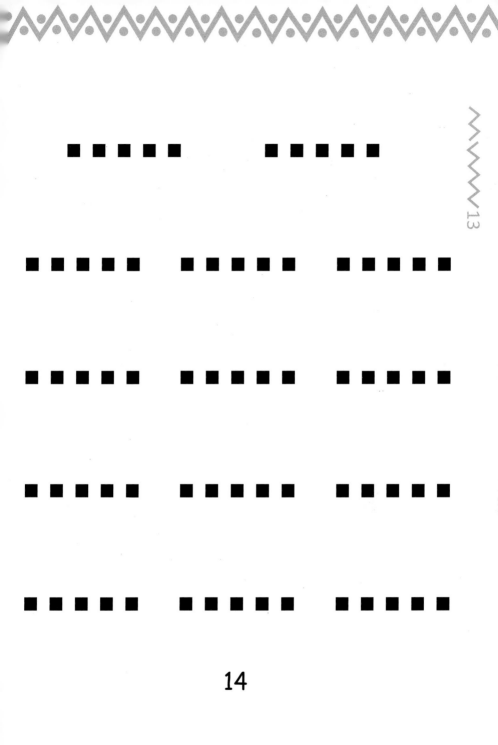

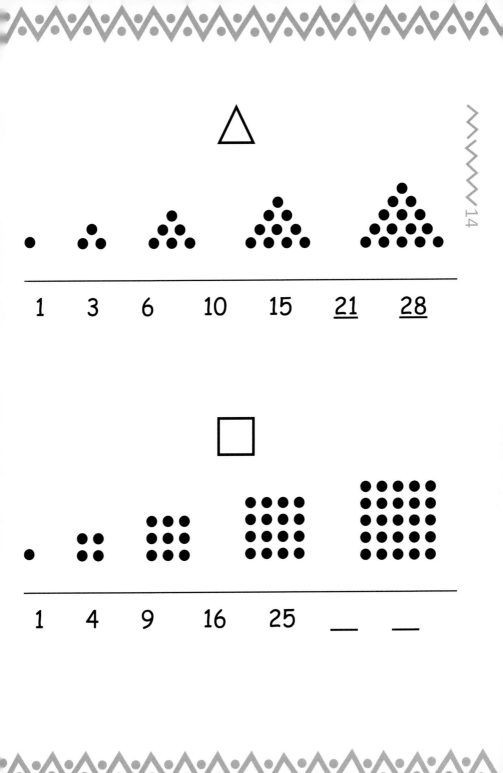

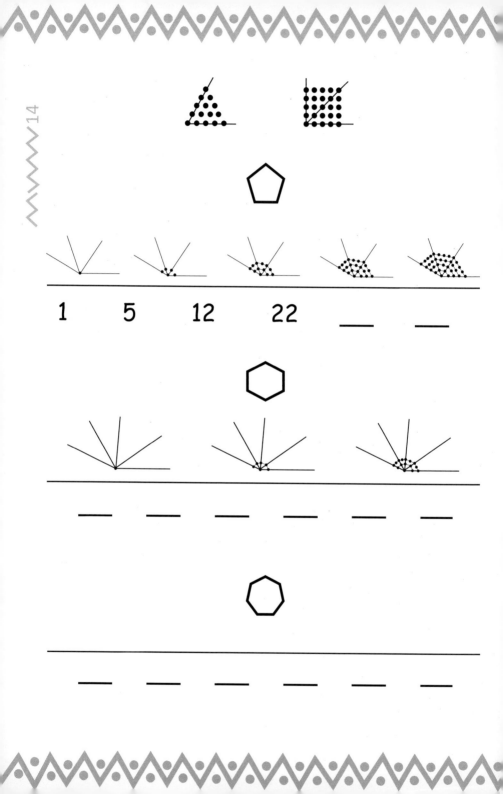

1   5   12   22   ___   ___

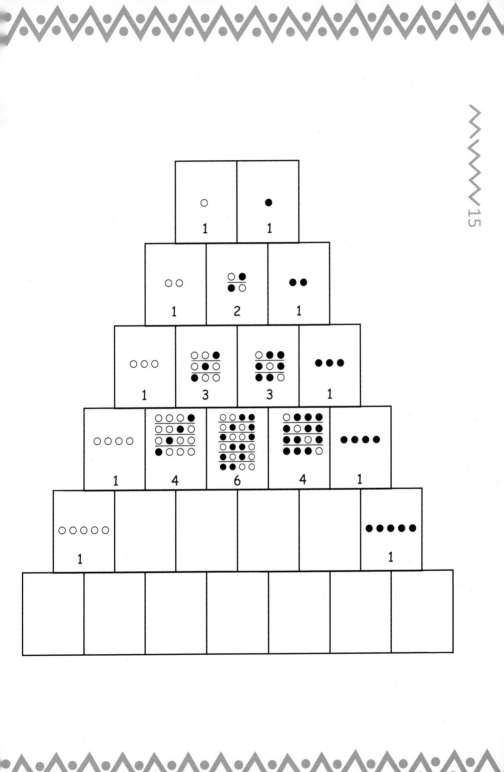

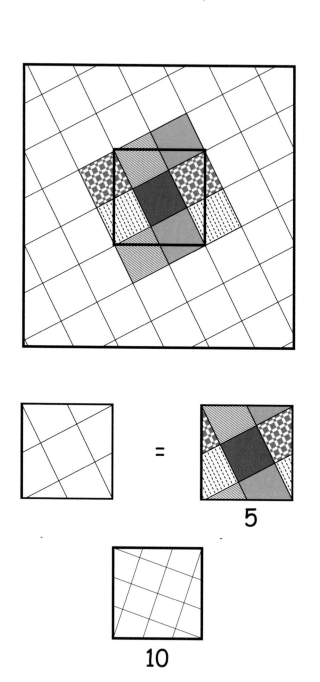

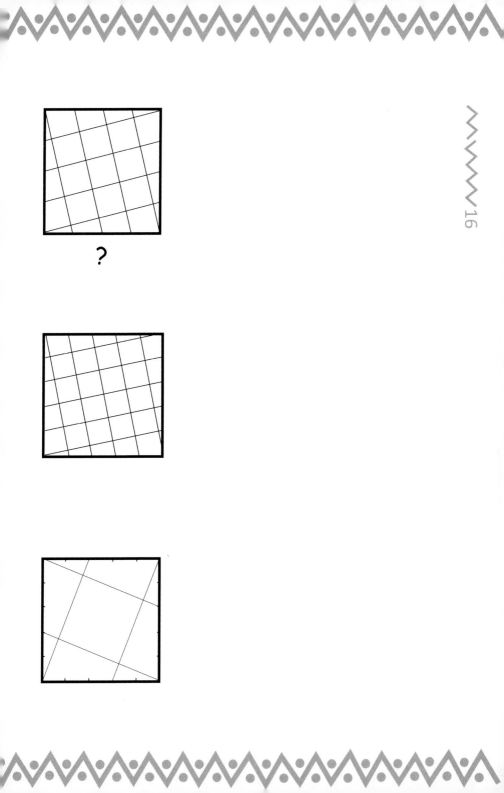

?

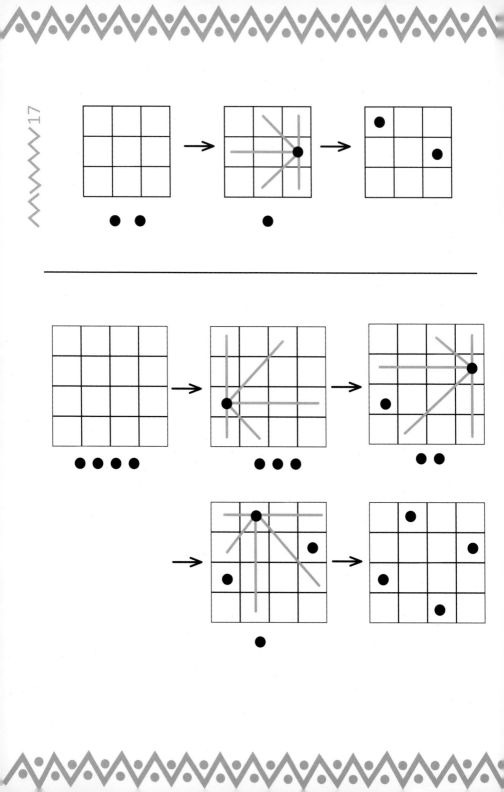

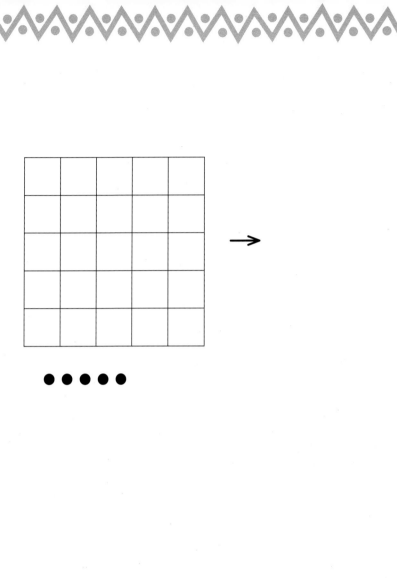

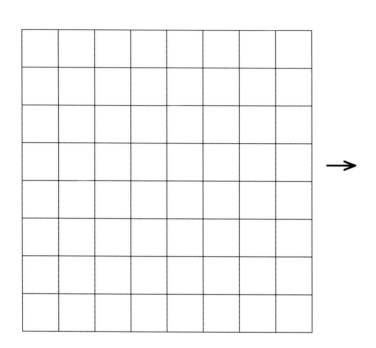

→

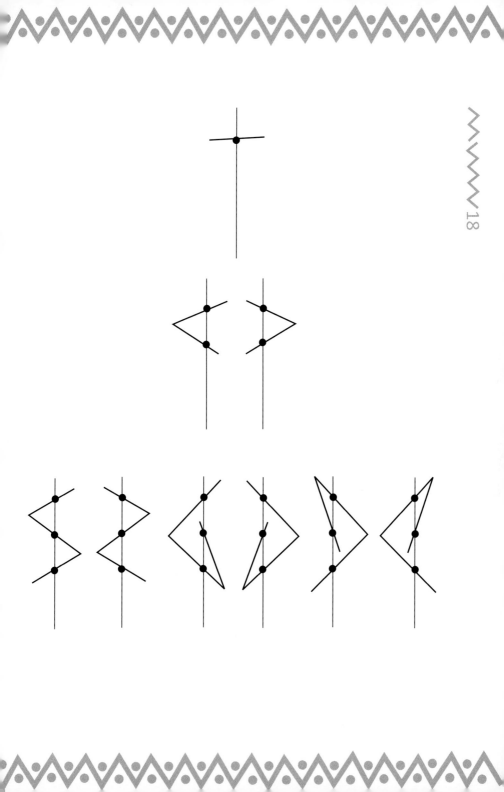

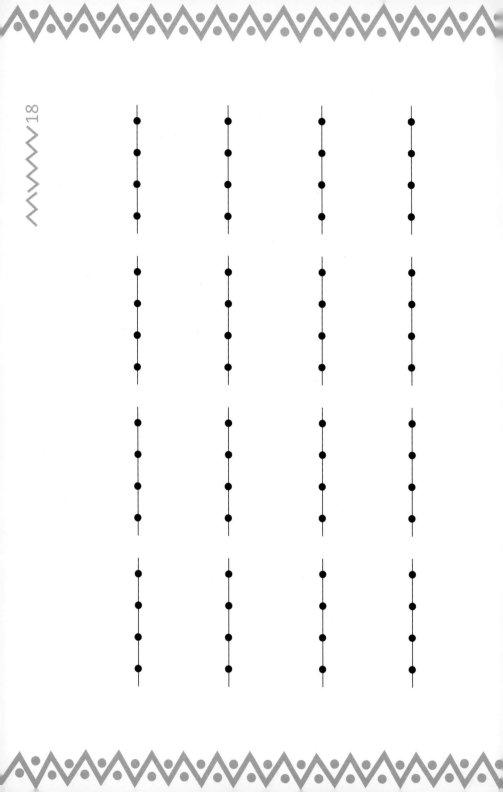

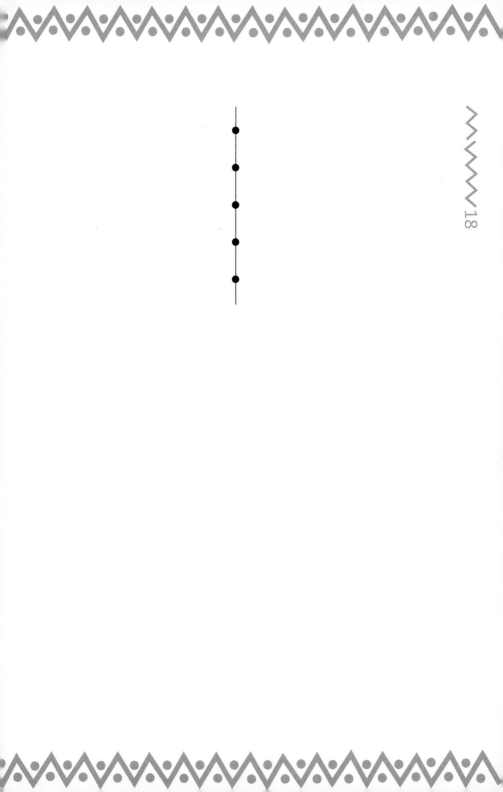

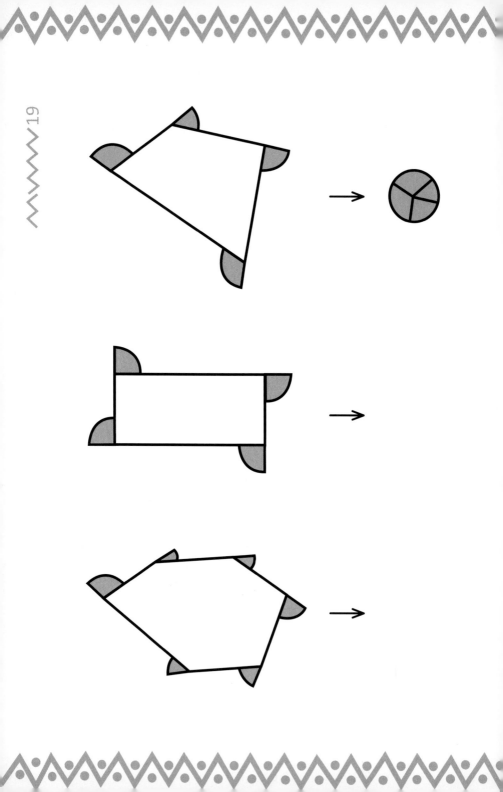

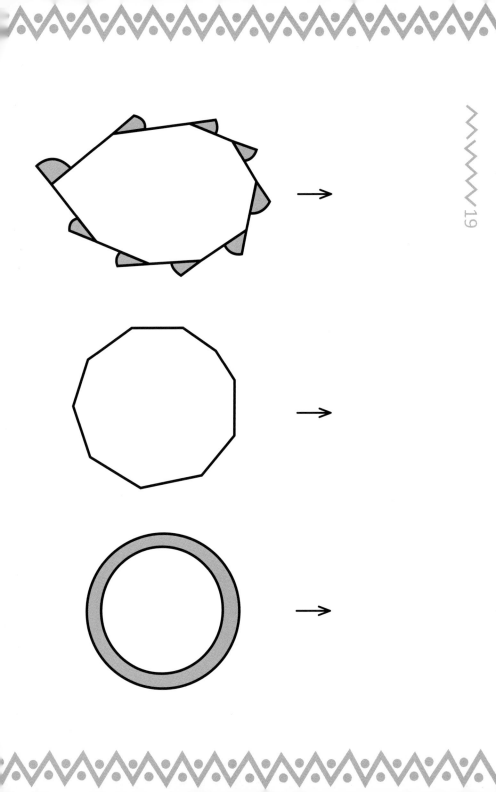

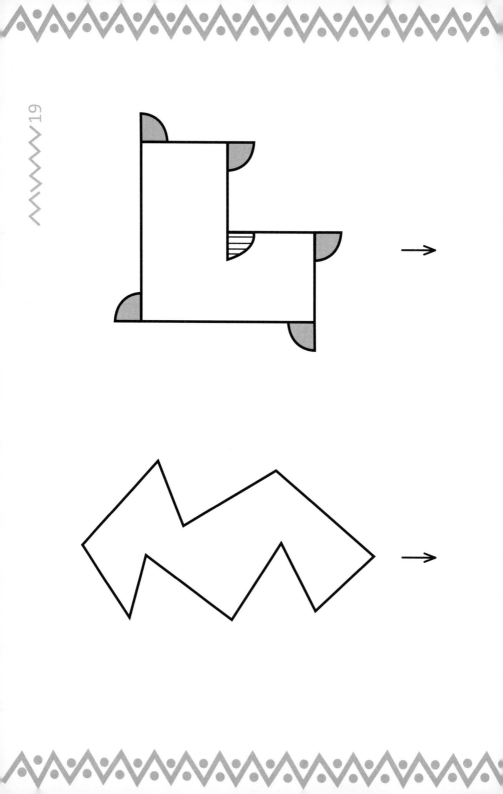

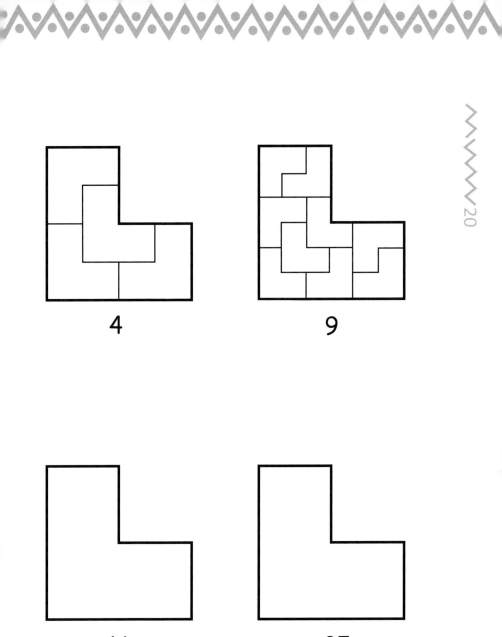

4

9

16

25

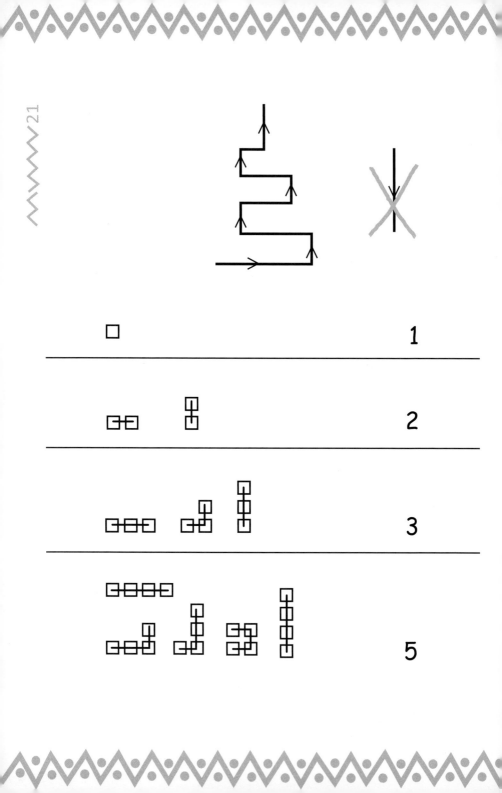

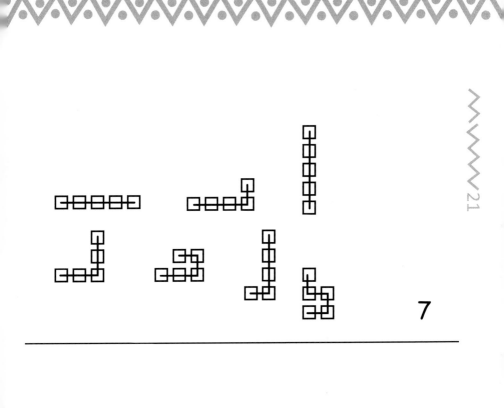

7

11

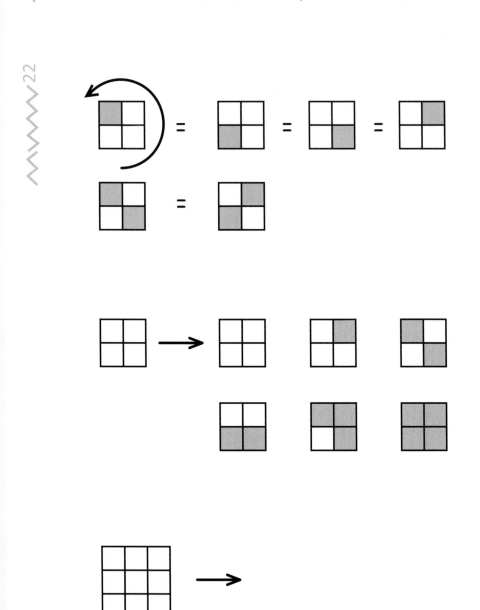

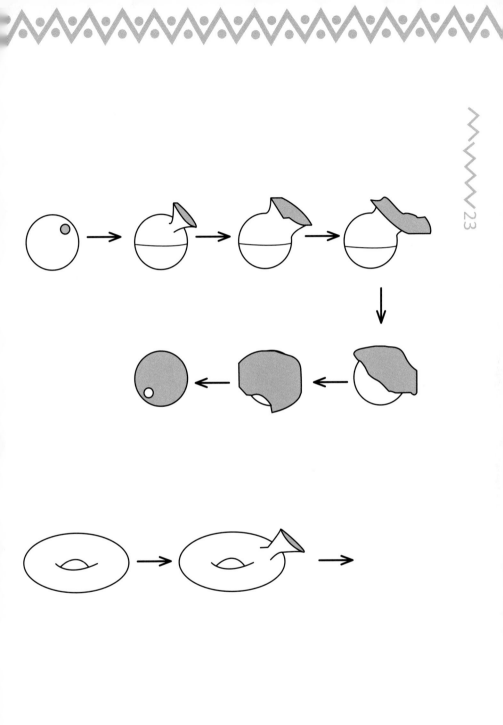

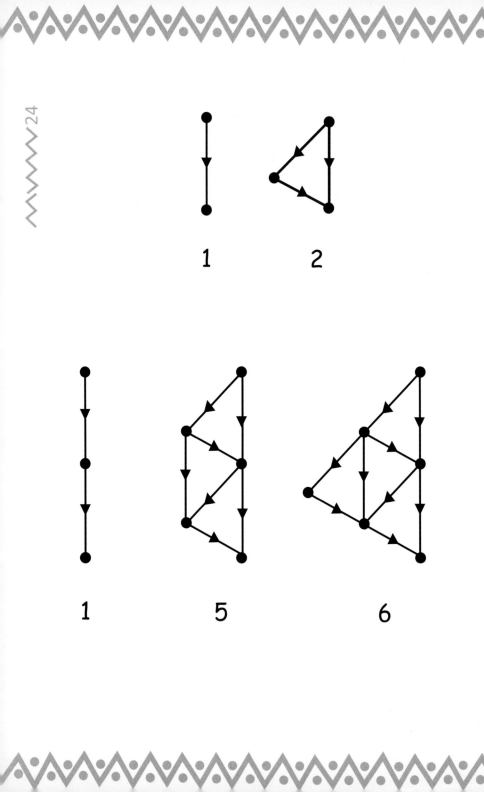

24

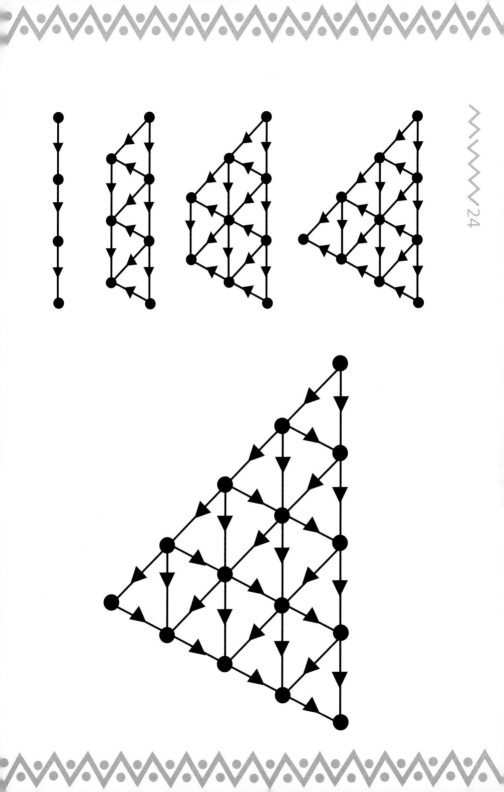

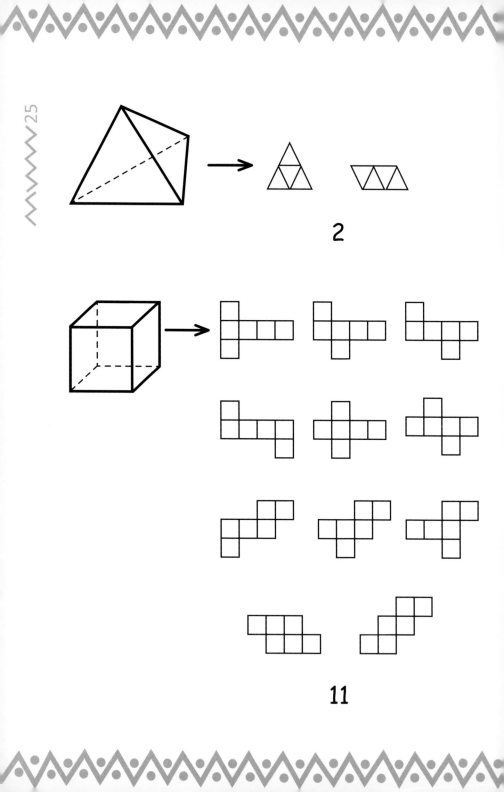

2

11

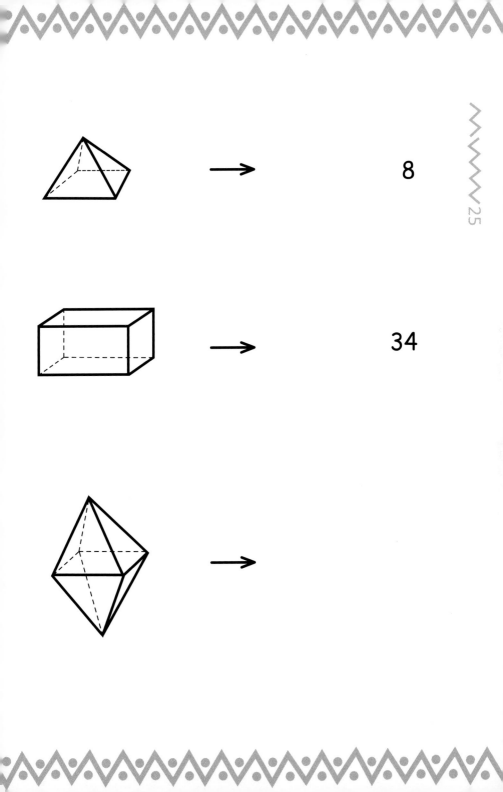

$\rightarrow$ 8

$\rightarrow$ 34

$\rightarrow$

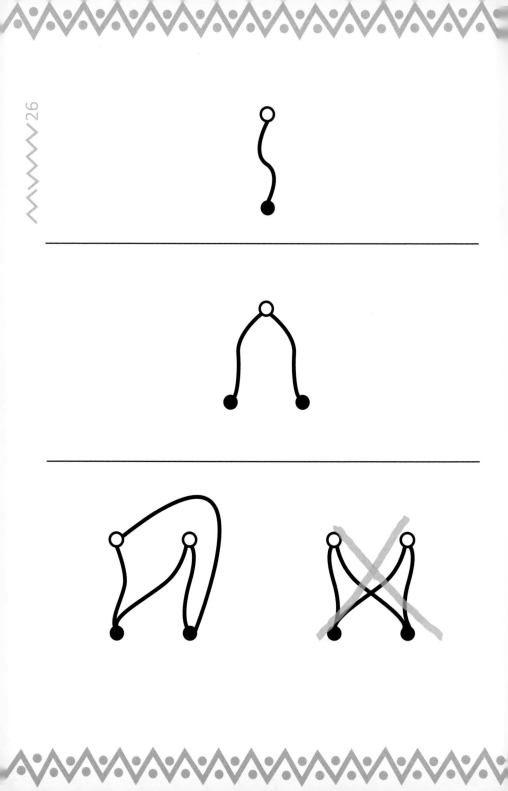

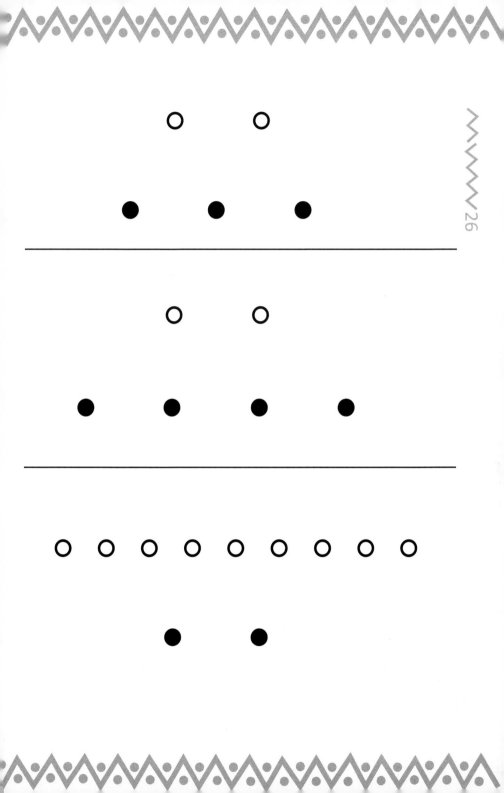

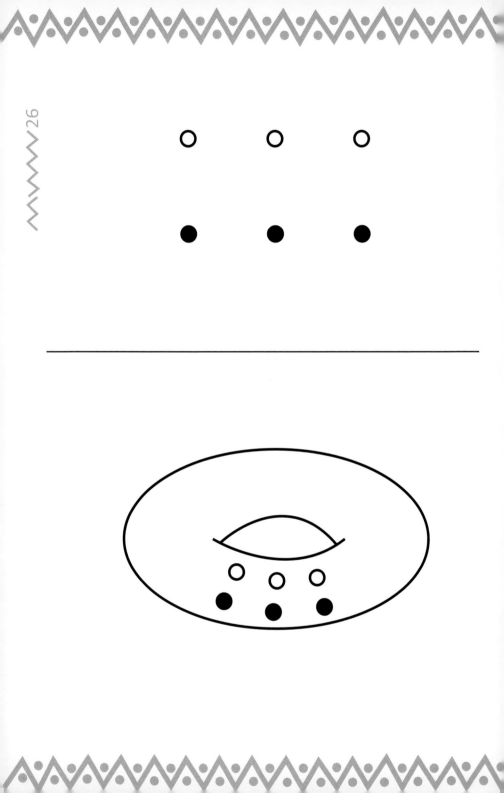

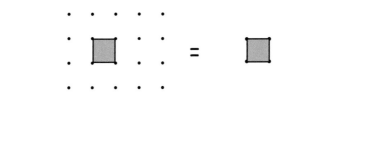

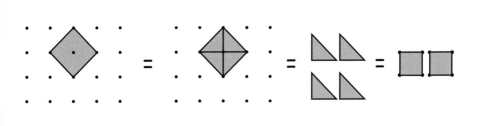

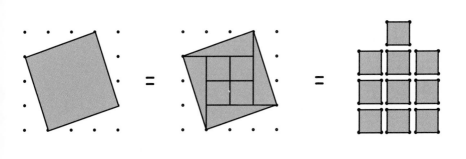

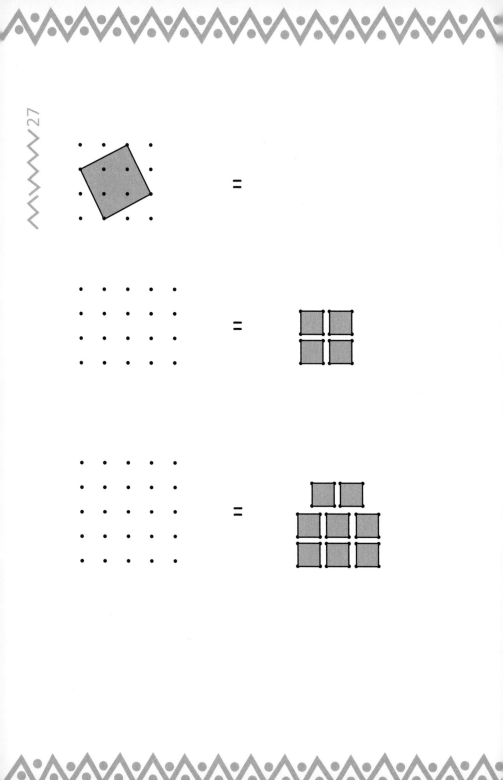

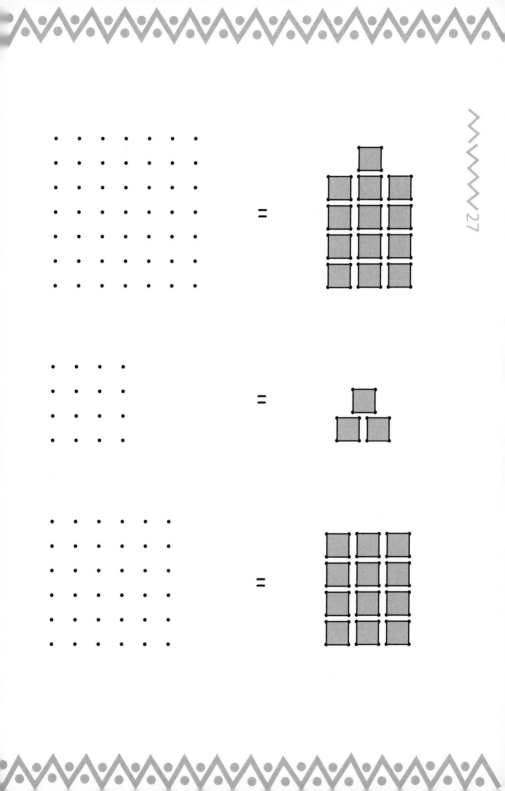

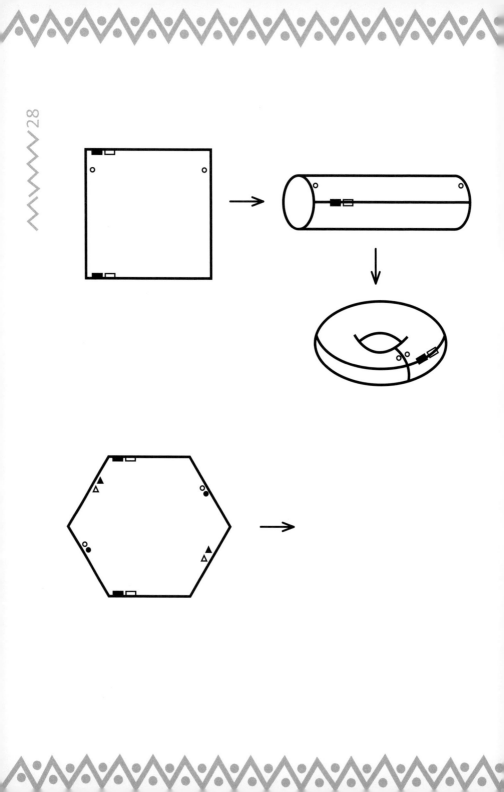

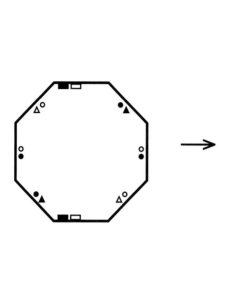

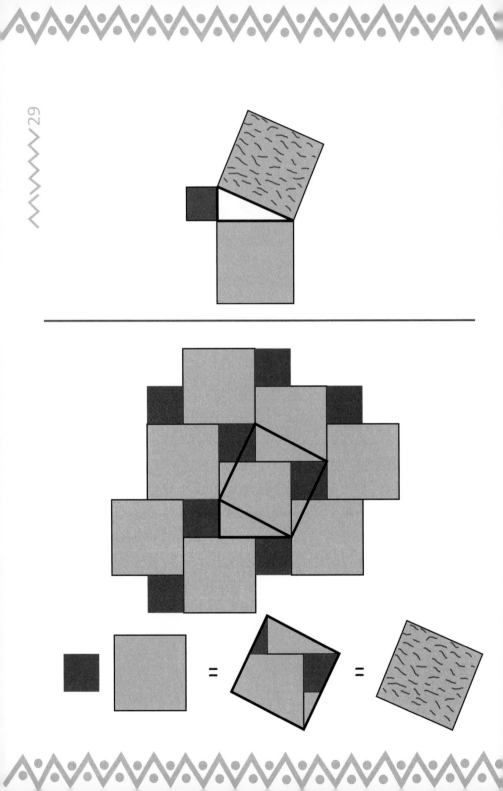

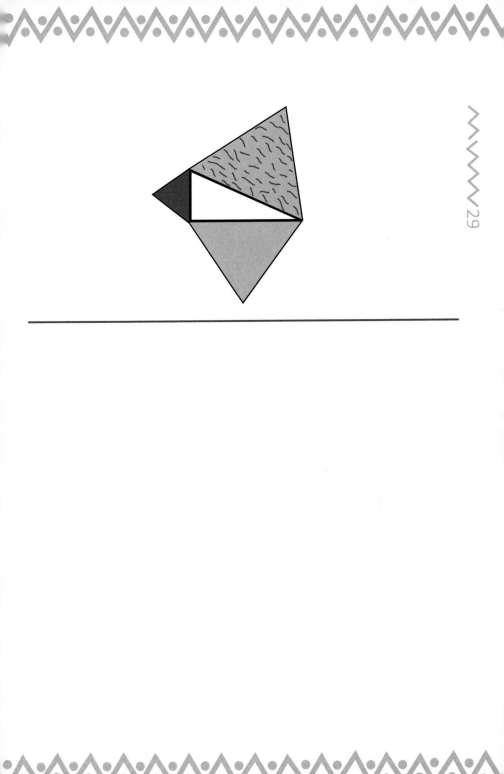

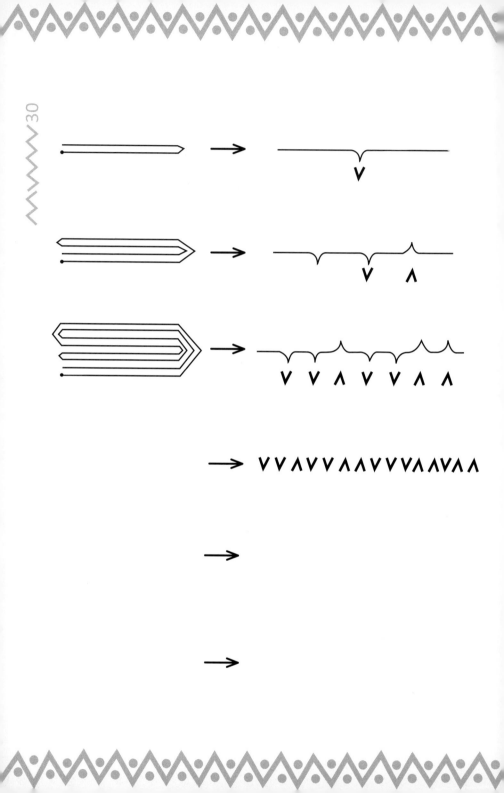

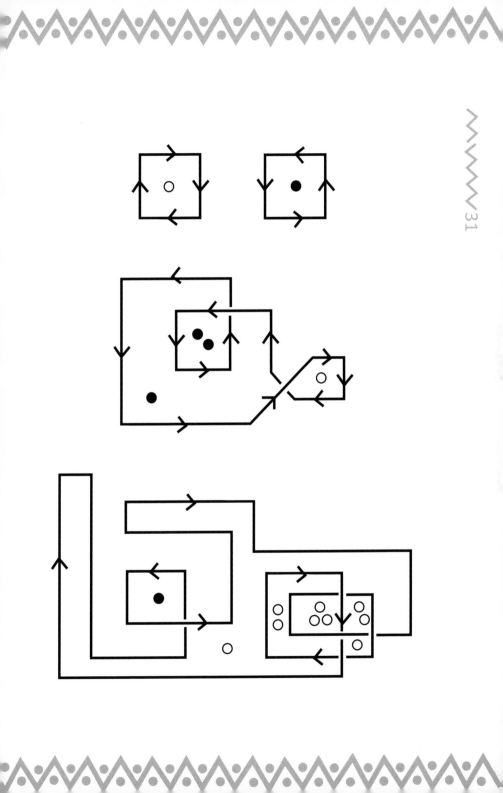

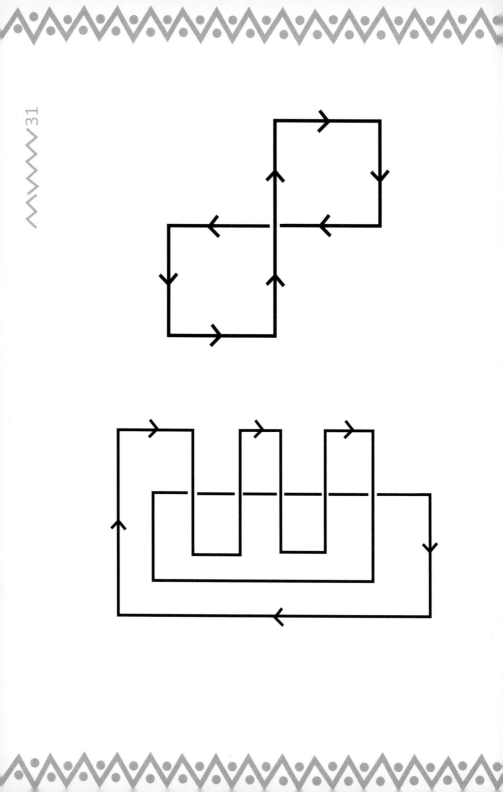

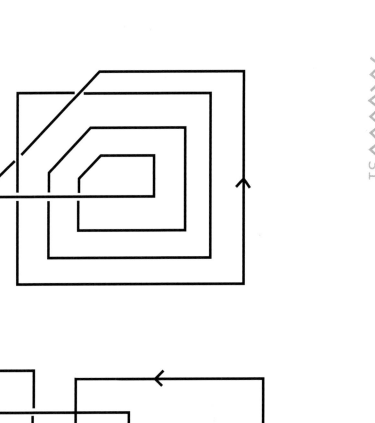

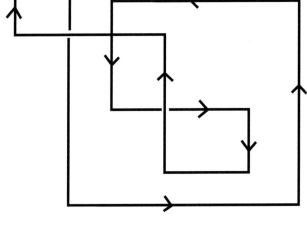

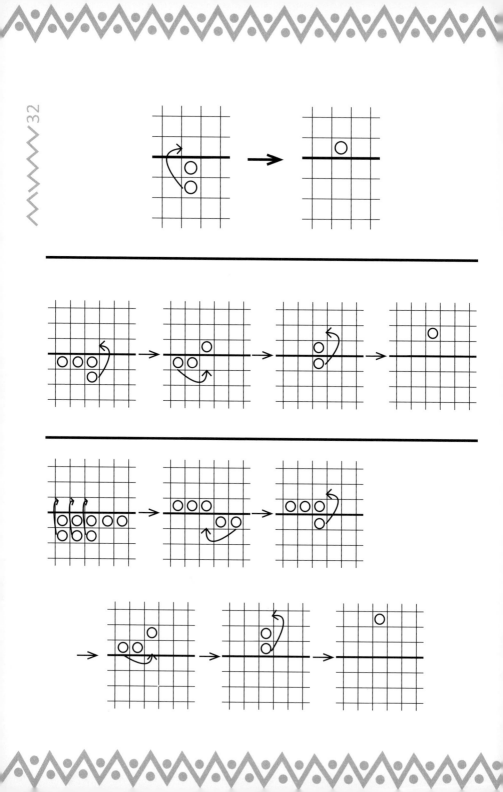

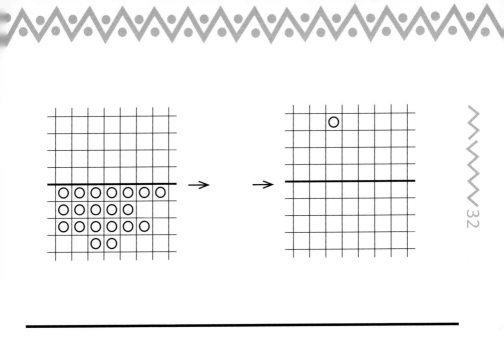

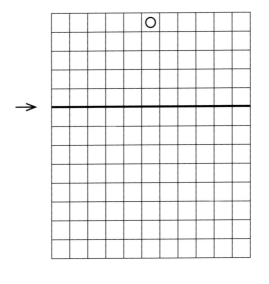

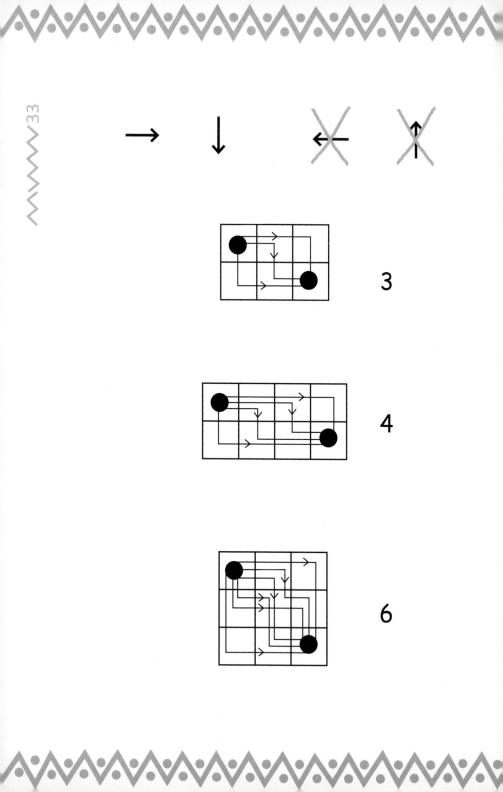

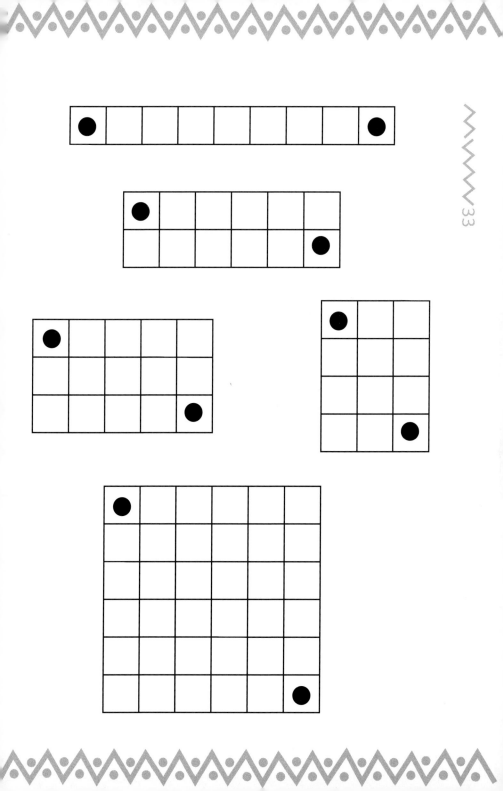

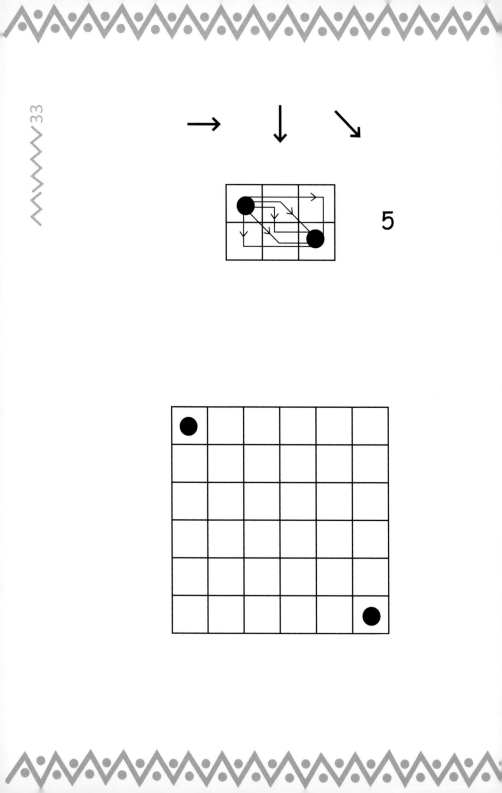

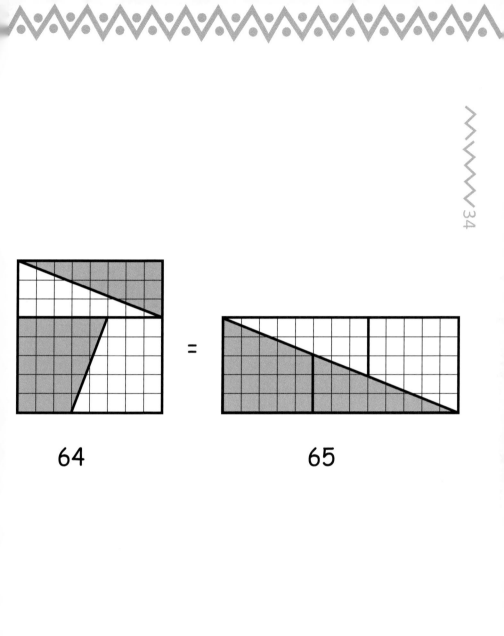

64   =   65

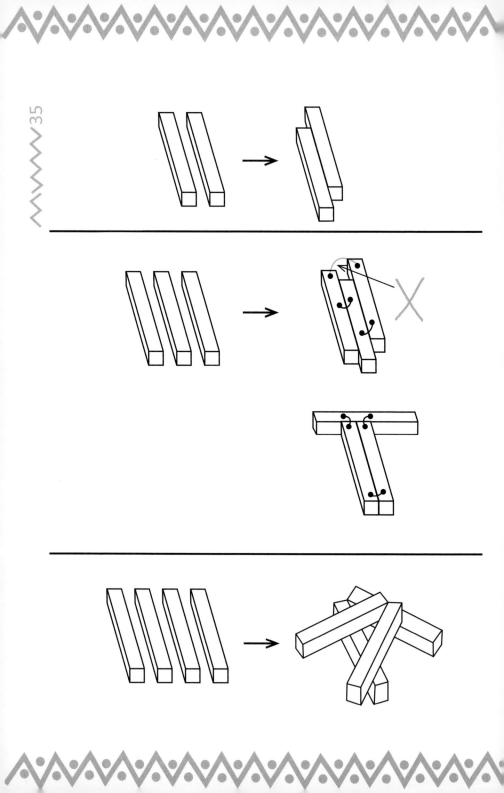

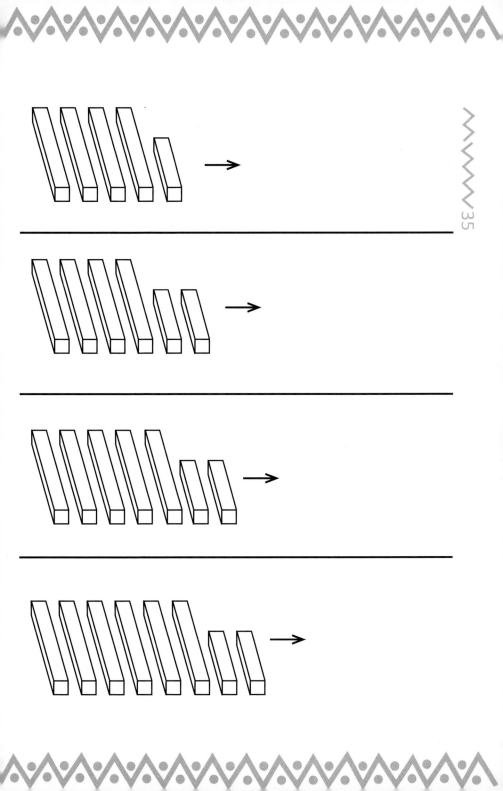

Techniques and Tips

... in case a few words are desired.

1. If you unfold these diagrams, how many tubes and how many loops do you get?

2. In how many ways can one stack blocks in rows, with each row of blocks aligned to the left? (Is the answer always a prime number?)

3. If two dots in a box "explode" away to become one dot, one place to the left, what happens if you place a large number of dots in a box? What if the rule changes so that three dots explode to become one dot one place the left, or three dots explode to become two dots one place to the left? See www.gdaymath.com/courses/ exploding-dots/ for the full glorious depth of this problem.

4. Can you tell whether or not a dot is inside a loop simply by counting the number of times a path from the outside to the dot crosses a boundary line?

5. In how many ways could you stack some bowling balls atop an initial bottom row?

6. In how many different ways can you break a row of squares into pieces?

7. If one draws straight lines across a page, into how many pieces do they divide the page? (Does the answer depend on how you choose to handle points of intersection?)

8. If a ball, starting at the bottom left corner of a rectangular grid, bounces off the sides at 45 degree angles, is it possible to predict into which corner the ball will fall? (Drawing becomes tedious after a while!)

9. In how many essentially different ways can one draw a set number of non-intersecting circles on a page?

10. Can you see the areas polygons drawn on a grid of dots?

11. In this game of solitaire a "move" consists of changing the color of any three distinct dots of your choice. Can you always turn a row of all white dots into a row of all grey dots?

12. If each symbol has a value, can you work out the indicated row sum and column sum from the information given?

13. In how many ways can you design arching pairs?

14. Arranging pebbles into different configurations gives the figurate numbers. We see the triangle numbers and the square numbers. Can you extend this idea and compute too the pentagonal numbers and beyond?

15. Count how many ways to place white dots and black dots in a row. Can you see how these counts are arranged to form a very famous triangular array of numbers?

16. Drawing a line from each corner of a square to the midpoint of an opposite side divides the square into five smaller squares: one central square and four squares of the same size each split into two parts. (A tiling pattern shows this.) If we draw lines from the third-way points of the sides, the square is then divided into the equivalent of ten small squares: four central ones and six each divided into two parts. (Can you use a tiling pattern to confirm this?) Why not push these observations further?

17. Can you place the dots on each board so that no two dots are in line vertically, horizontally, or diagonally?

18. Can you count the number of ways a straw could be folded so as to pass through each dot given on a vertical line? (Equivalently: In how many different ways could you fold a strip of paper n units long into a unit-long segment, n layers thick?)

19. Do the exterior angles of a polygon always add to a full turn?

20. Can you divide an L-shape (three unit squares pasted side-by-side) into the indicated number of congruent smaller L-shapes?

21. In how many ways can one stack a collection of blocks if the blocks must follow a path that can be seen as going right, up, left, up, right, up, left, up, ...?

22. There are six ways to colour the cells of a square divided into four, assuming rotations and a reflections of the same design are deemed equivalent. How many essentially distinct colour designs are there of nine cells?

23. Turning a hollow rubber ball inside out yields an object of the same shape. What shaped object results from turning a hollow rubber donut inside out? (Can it even be everted?)

24. How many different routes are there from top to bottom following the one-way streets?

25. There are two different paper cut-out designs that fold to make a tetrahedron. There are 11 different paper cut-out designs that fold to make a cube. Can you count designs for making other three-dimensional shapes? (No overlapping paper allowed.)

26. Can you connect each white dot to each black dot with curvy lines that do not intersect?

27. Determine the permissible areas of tilted squares drawn on a grid of dots.

28. Gluing opposite edges of a flexible square sheet of material makes a donut. What shape is produced by gluing together opposite edges of other polygonal sheets of material?

29. By tiling a floor with square tiles of two sizes, we see a proof of the famous Pythagorean Theorem: If squares are drawn on the sides of a right triangle, then the largest square has area the sum of the areas of the two smaller squares. The analogous result is true too for equilateral triangles drawn on the sides of a right triangle. Can this be seen purely geometrically?

30. If you fold a strip of paper repeatedly in half, always picking up the right end and folding it over to the left end, what interesting patterns do you see in the creases that result? Can you explain the patterns you see?

31. Each region is labelled with some coloured dots. They count the number of turns the path makes around that region. Can you see how?

32. Leapfrog one coin over a neighbouring coin into an empty space and remove the coin that was jumped over. With these moves can you get a coin a given number of rows beyond a fixed horizontal line?

33. How many paths are there between dots if one can only make single right steps and single down steps?

34. Surely this picture can't be right? The four pieces are identical in shape, yet seem to make rectangles of two different areas! What is going on?

35. Can you stack blocks of wood so that each piece is in contact with each and every other piece?

There are solutions to the puzzles - and a wealth of other wonderful mathematical material – on the author's website: **www.jamestanton.com**.

More from Tarquin

If you've enjoyed this book then you'll enjoy its sister book Without Words, too – and you might be interested in the Without Words Poster Set that accompanies the books – full details and images on www.tarquingroup.com.

There you will also find puzzles, books, dice and posters all with a mathematical theme. Our products are used worldwide by teachers, tutors, parents and recreational mathematicians.

Tarquin has been publishing for more than 35 years and we have something for everyone!

www.tarquingroup.com

For a full catalogue email us at info@tarquingroup.com or write to us at:

Tarquin
Suite 74, 17 Holywell Hill
St Albans
AL1 1DT
United Kingdom

Our books and many other products are also available on Amazon and many other retailers worldwide.